Math Mammoth
Foundational Word Problems

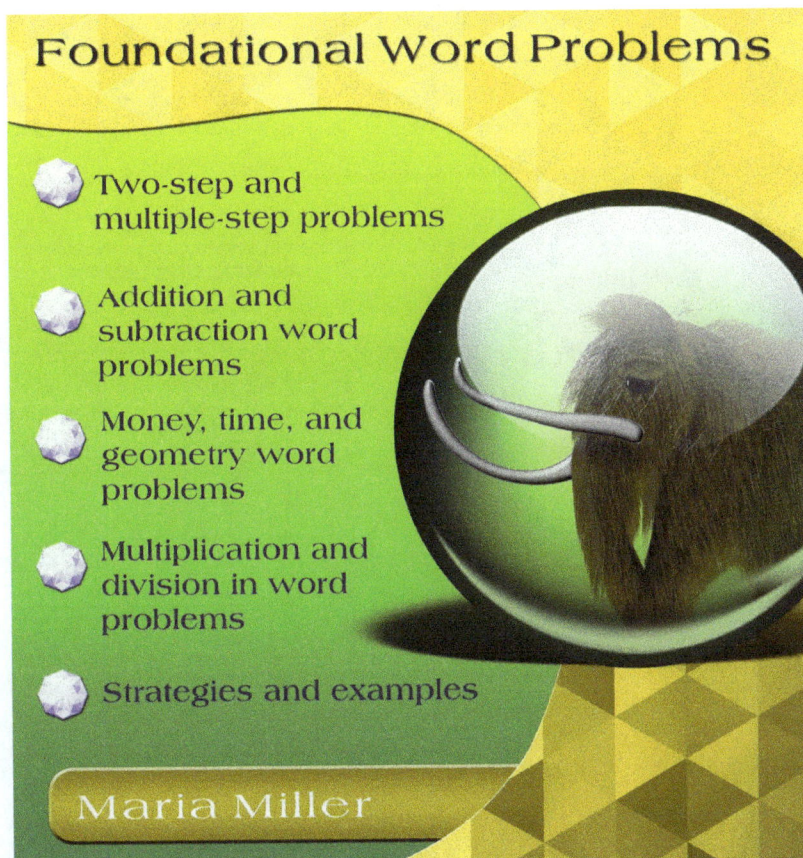

Foundational Word Problems

- Two-step and multiple-step problems
- Addition and subtraction word problems
- Money, time, and geometry word problems
- Multiplication and division in word problems
- Strategies and examples

Maria Miller

By Maria Miller

Contents

Introduction

Math Mammoth Foundational Word Problems is intended for students in 3rd grade and above that need more practice with word problems. I especially aim to help students in 4th-7th grades who have trouble with word problems and who need to "go back to the beginning", so to speak, to learn to solve them.

The problems are generally based on a knowledge of third grade math, such as regrouping in addition and subtraction, the multiplication tables, concept of division, and calculating change. The problems do not require long multiplication, long division, or calculations with fractions or decimals (with the exception of money amounts).

The worksheets have examples and some instruction about common strategies for word problems. Students are often encouraged to write down the calculations they do. They are also encouraged to make a drawing or a sketch of the situation in the problem, to help them understand the situation and to see what math operation is involved.

Overall, most of the problems require two or more steps; however, some problem sets focus on differentiating what math operation should be used, and those only require a single step.

Several problem sets are designed to help students to master problems that involve equal-size groups. These questions either ask for the total (and are thus solved by multiplying), or they ask for the size of the group or for the quantity of the groups (one can write a missing number multiplication sentence, such as $3 \times \underline{\quad} = 18$ or use division).

The book also includes word problems with fractions, with large numbers, with money, geometry word problems, and challenges in the form of puzzle corners.

I hope you will find that the book meets your needs!

I wish you success with teaching math!

Maria Miller

Solving Word Problems

The key to solving word problems is to see what math operation(s) match the situation.

Read the problem carefully, and imagine in your mind what is happening. You can also draw a picture!

Example. Here, the solution to the problem is written out for you. Circle the correct word and fill in the missing parts.

A laptop cost $650, but Eli got a $40 discount. What was the final price?

Eli paid $40 _less / more_ than $650.

It means we _subtract / add_: $650 ☐ $40 = _____

The final price was _____.

1. Circle the addition and / or subtraction sentence that matches the problem. Solve.

a. Maggie has 62 glitter pens. Of them, 15 don't work, and the rest are good. How many glitter pens do work?

$62 + 15 =$ _?_

$62 +$ _?_ $= 15$

$15 +$ _?_ $= 62$

$15 - 62 =$ _?_

b. Cash rode 148 km in a bus, and then walked 4 km to Grandma's house. How many kilometers did he travel?

$148 +$ _?_ $= 4$

$148 + 4 =$ _?_

$4 +$ _?_ $= 148$

$148 -$ _?_ $= 4$

c. The temperature outside is 46 degrees Fahrenheit, and inside it is 74 degrees. What is the difference in the temperatures?

$74 + 46 =$ _?_

$74 +$ _?_ $= 46$

_?_ $- 74 = 46$

$74 -$ _?_ $= 46$

d. Zoe needs 54 pins for a sewing project. She only has 44. How many more does she need?

$54 + 44 =$ _?_

$44 -$ _?_ $= 54$

$54 +$ _?_ $= 44$

$54 - 44 =$ _?_

2. Solve. On the empty line, write an addition or a subtraction sentence (or several) to match each problem.

a. An airplane has 143 seats. Of these, 46 are window seats
How many seats on the airplane are *not* window seats?

_____

_____ seats are not window seats.

b. One hundred seventy-five people were called to a meeting. Four couldn't come and nine others didn't show up. How many *did* come for the meeting?

_____

_____ people did come to the meeting.

c. During a year, a town had 104 rainy days. How many days did it not rain that year? (One year has 365 days.)

_____

It did not rain on _____ days.

d. Luca has 14 toy cars, Elliott has 25, and Owen has four more than Luca. How many more cars does Elliott have than Owen?

_____

_____

Elliott has _____ more cars than Owen.

e. Blake has saved 74 dollars. How much does he still need to buy a bicycle that costs $80?

_____

Blake still needs $_____ .

Word Problems: Mental Math

Example. Aurora bought some flowering plants and planted eleven of them. Now she has four left to plant. How many plants did she buy?

Drawing the situation can help, as shown on the right. Just a rough sketch is enough!

Then think what math operation fits the situation, and write a number sentence (or several). You can use x or some other letter for what is not known.

Here, we don't know how many plants Aurora bought, so that is the unknown, or x. She bought x plants, planted 11, and has 4 left. That is a subtraction situation: $x - 11 = 4$

Another number sentence you could write is: $x = 11 + 4$.

Either way, the answer is: Aurora bought 15 plants.

1. Solve. Make a drawing of the situation to help you!

a. Dad ate 11 cookies, Mom ate six, little sister ate three, and there are still seven left. How many cookies were there originally?

b. Twelve children were playing on the playground. Five of them left. Then, ten more children came. How many are playing on the playground now?

c. Marcus has 24 marbles and Jayden has 16. How many do the boys have together?

Could they share all of those marbles evenly? If yes, how many would each boy get?

2. Match the correct number sentence(s) with the problem.

a. Claire is 131 cm tall and Molly is 4 cm shorter than her. How tall is Molly?	$131 + 4 = x$ $\qquad$ $131 + x = 4$ $131 - x = 4$ $\qquad$ $131 - 4 = x$
b. Three months ago, Bentley was 133 cm. He has grown 6 cm since then. How tall is he now?	$133 + 6 = B$ $\qquad$ $B + 6 = 133$ $6 + B = 133$ $\qquad$ $B - 6 = 133$

3. Solve the problems. Can you draw something to picture the situation?

a. One day, a migrating whale swam 50 miles. Then it swam 15 more miles, and then 30 more miles. Then the day was over. What was the total number of miles the whale swam that day?

b. You had $40. You purchased two bouquets of roses for $8 each. How much do you have left after the purchase?

4. Write an addition or a subtraction for each problem. Use a letter for the unknown number.

a. Holly earned $5 for raking the yard and $8 for weeding. She had already saved $35. How much money does she have now? _____ Answer: _____	**b.** Natalie had $20. She bought a gift and now she has $9 left. How much did the gift cost? _____ Answer: _____
c. A tablet that costs $199 has a sales tax of $12 added to the price. What is the final price? _____ Answer: _____	**d.** Isabella bought the tablet and paid with $250. What was her change? _____ Answer: _____

Multiple Steps — Addition and Subtraction

Example. The Blackwood family visited a beach that is 142 km away from their home. On their way home, they stopped to rest when they still had 80 km to go. How long a distance have they driven in total, so far?

A simple **drawing can help you "see" the situation.** They drove 142 km all the way to the beach. When they headed home, they drove an unknown distance, and had 80 km left.

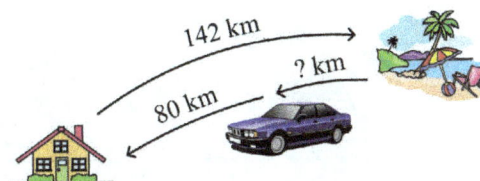

This problem takes two calculations to solve.

(1) First we find the unknown distance x.

(2) Then we find the total distance they have traveled so far.

Example of student work:
(calculations written down)

$x = 142$ km $- 80$ km $= 62$ km

Total distance:
142 km $+ 62$ km $= 204$ km

When solving word problems, **write down each calculation you do**, even if you could do it in your head. See the example above: it took two calculations, and both are written down. The second one also has the words "Total distance" to signify what the calculation is for.

All of this helps you organize your thinking. It also helps you (and others) to *easily* read what you did, and thus check your work. And it's not only the teacher that needs to read and understand what you did — you yourself need to do it, so you can check your work!

Monitoring your own work, checking if what you did is correct, is REALLY important in real life! And I don't mean just math, but everything! ☺

1. Solve. You can draw to help! Draw items with price tags, for example. Write down each calculation.

One watch costs $25 and another costs $8 less. Ian buys two of the cheaper watches. Find the total cost.

2. Solve. Write down each calculation you do, even if you could do it in your head.

a. Dante had 64 golf balls and Jace had 31. Then Dante lost nine of his. How many *more* balls does Dante have now than Jace?

b. Brady has saved $270. He wants to buy a camera for $320 and its batteries for $20. How much does he still need to save?

c. Nadia wants to buy a stuffed elephant for $11, a stuffed lion for $23, and a stuffed pony for $14. She still needs to save $19 more, before she can buy them. How much money does she have now?

d. Grandma Elaine is sewing a quilt. She is planning for there to be 167 squares with plain colors and 185 squares with patterns. She sewed a total of 138 squares onto the quilt last month. How many squares does she have left to sew?

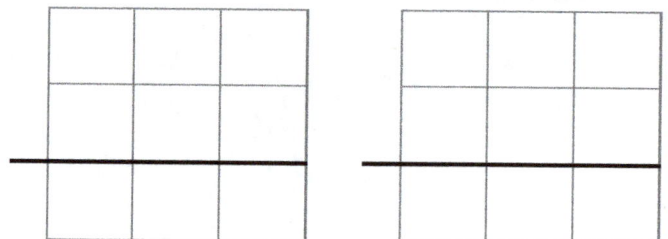

3. Solve. Write down each calculation you do, even if you could do it in your head.

a. Mr. Gale took two 25-kg barrels of mangoes, a 30-kg sack of lemons, and a crate of plums weighing 40 kg to the market. There he sold 87 kg worth of his produce. How many kilograms of fruit does he have left?

b. The Howard family's house has three leaky sinks. Last year, it cost $115 to fix a leaky sink, but this year it costs $8 more per sink. How much will it cost the Howards to fix their leaky sinks this year?

c. Lakeside School had 338 students, and Elk Valley School had 727. Then 116 students moved from Elk Valley School to Lakeside School. Which school has more students now?

How many more?

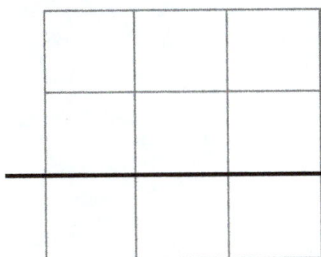 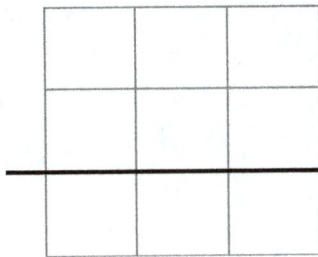 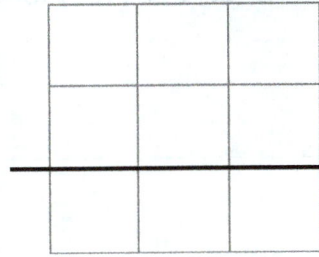

More Addition and Subtraction Word Problems

1. Solve. You can draw to help! Write down each calculation you do.

a. An exercise mat costs $56. Kara has $27, and then her grandma gave her $15 more. How much more money does she still need before she can buy it?

_____

_____

b. Pencils are packaged in boxes of 64. Aliyah has three boxes. She took 13 pencils out of one of the boxes. How many pencils are in the boxes now?

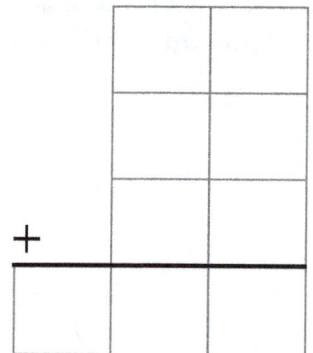

_____

_____

c. One laptop costs $429 and another costs $30 less than the first. Camille buys both. How much do they cost together?

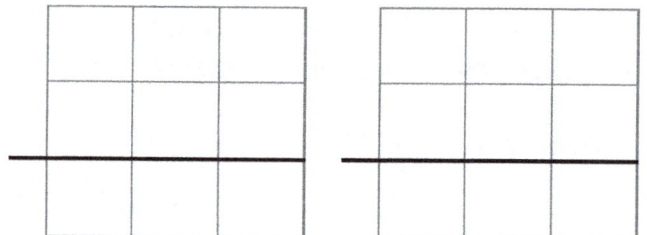

_____

_____

d. Ken earned $95 each week for three weeks. Then he paid his parents $130 to help with the cost of food. How much does he have left for himself?

_____

_____

3. Solve.

a. One tree grows 355 limes and another grows 52 more limes than that. How many limes did <u>both</u> trees grow?

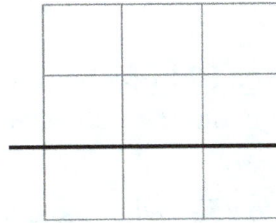

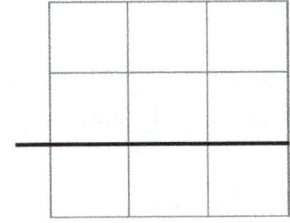

_____

_____

_____

b. A big box contains 400 sweaters in different colors. There are 75 green sweaters, 115 yellow ones, and the rest are blue. How many are blue?

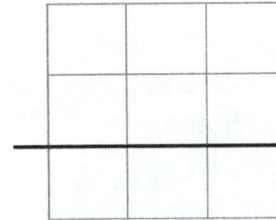

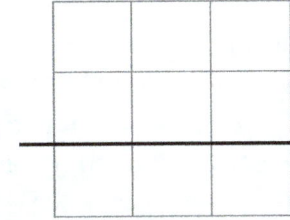

_____

_____

_____

4. Now it's your turn! Write a word problem to match each equation.

a. $14 + \triangle = 31$

b. $\triangle - 32 = 105$

Extra Numbers in Word Problems

Be careful! Sometimes, word problems have extra numbers.

Example. Alyssa is on page 224 of a book that has 380 pages. Then, she reads 29 more pages, and is on page 253. How many pages are left to read?

The numbers that matter are 253 (what page she is on) and 380 (the total number of pages). The underlined numbers in the problem are not necessary.

We can write an addition: $253 + x = 380$. This is solved by subtracting $380 - 253 = 127$.

1. Solve the problems.

a. Theo has saved $332. The guitar he wants to buy costs $489. He works for $15 an hour. How much more does he need to save to buy it?	**b.** A calf is born weighing 43 kg. A year later, it has gained 286 kg. Another year later, it has gained another 121 kg. How much does it weigh now?
c. One laptop costs $369 and another costs $45 more. Xander buys two of the cheaper laptops. Find the total cost.	**d.** Finley is starting a farm. She drives 28 km and buys 25 chickens. The next month, she buys 13 geese and 18 ducks. It costs her $11 per bird. How many birds did Finley buy?

3. Solve the word problems. Write a number sentence or several for each problem.

a. During the summer, Bianca, Laree, and
Reuben worked cleaning a hospital. Laree
earned $59 less than Bianca, Bianca earned
$137 more than Reuben, and Reuben
earned $785. Bianca worked a total of
61 hours. How much did she earn?

_____

_____

She earned _____ .

b. Silas mixed together pieces from two 164-piece puzzles and
from one 56-piece puzzle. How many pieces are in the pile?

_____

_____

There are _____ pieces in the pile.

c. In a single month, Leo the puppy chews up 41 bones and three shoes. The next month,
he chews up eight fewer bones than the first month, two more shoes than the first month,
and one carpet. What was the total number of bones and shoes Leo chewed up?

_____

_____

Leo chewed up _____ bones and shoes in total.

d. A farmer has 92 ewes and three rams. Then the ewes
give birth to 163 lambs. After seven months, he sells
eight ewes and 139 lambs.
How many sheep does he have now?

_____

_____

Now he has _____ sheep.

Rounding

Sometimes you don't need to know the exact answer, and you can simply estimate the answer, using rounded numbers.

You can also use rounded numbers and estimation to help check calculations. If the estimated answer is not close to the exact answer there is probably an error.

Example. Tessa wants to buy a bicycle for $139. She has $58. How much more does she need before she can buy the bicycle?

To estimate the answer, we round the price of the bicycle to $140 and the amount Tessa has to $60. So, she needs about $80 more to buy the bicycle.

But the subtraction on the right shows $91 as the answer. This is a bit far from $80. Can you find the error?

$$\begin{array}{r} 13 \\ \cancel{1}\ \cancel{3}\ 9 \\ -\quad 5\ 8 \\ \hline 9\ 1 \end{array}$$

1. Round some or all of the numbers to the nearest ten and estimate. Then find the exact answer. If the exact answer and the estimate are not close, check your work, because you have probably made an error somewhere.

a. A farmer has 139 chickens and 16 turkeys.
About how many birds does he have?

Estimation: _____

Answer: _____

b. About how much would one $38 DVD and two $29 DVDs cost together?

Estimation: _____

Answer: _____

c. Sienna went to town with $776 and bought a camera. Then she had $147 left. About how much did the camera cost?

Estimation: _____

Answer: _____

2. Estimate, using rounded numbers, the total distance all the way around this triangular path in the forest. Also, find the real distance by adding the original numbers in the grid.

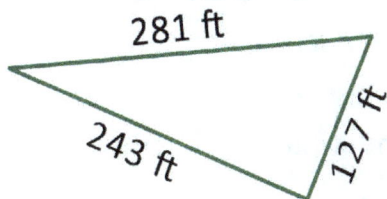

281 ft

243 ft

127 ft

Estimate: _____ ft

3. Round to the nearest ten, then solve using mental math. In these problems, you don't need to find the exact answer. Remember: numbers ending in 5 are always rounded up.

a. Marissa is driving to her grandparents' house 233 kilometers away.
About how many kilometers is it there and back?

b. Andreas goes on three shopping trips during the week. He spends $129 the first time, $115 the second time, and $210 the third time. About how much did he spend?

c. During the year 2022, Swampville had 178 rainy days.
About how many days did it not rain? (One year has 365 days.)

d. Vivian is on page 292 out of 364 in a book. About how many pages does she have left?

e. Which is the better deal: two phones that cost $369 and $512, or two phones that cost $445 each?

Money Situations

Example. *Flashlights cost $13 each. How many can Derek buy if he has $50? How much will he have left?*

To find how many times 13 can fit inside 50, you can add $13 multiple times until you've found the answer:

$$\$13 \quad + \quad \$13 \quad + \quad \$13 \quad + \quad \$13$$

+ = $26 + = $39 + = $52

Derek can buy three flashlights that cost $13, which would cost $39. Then we subtract $50 − $39 to find he would have $11 left over.

1. Solve.

a. How many backpacks that cost $34 each can you buy with $100?

How much money will you have left?

b. Mugs cost $8 each. How many can you buy if you have $40?

How much money will you have left?

c. Caleb wants to buy a tablet for $169. He has $10, and he earns $40 each week. In how many weeks will he have enough to buy the tablet?

2. Round the prices to estimate your total. Then find the exact total. Remember: if your answer is far away from your estimation, there is probably an error.

a. You buy headphones for $16.50 and a watch for $38.20.

Estimated total $_____

Exact total $_____

b. You buy an ice maker for $155.45 and a set of kitchen utensils for $37.99.

Estimated total $_____

Exact total $_____

3. Solve the problems.

| $7.90 | $5.25 | $2.40 | $28.50 | $36.90 |

a. How many bottles of glue can Ernest buy with $10?

b. Daniel bought a toaster and paid with a $50 bill. He received $21.50 as change Was that correct?

c. Can Carol buy three computer mice and a microscope with $70? Use estimation and mental math.

Multiplication in Word Problems

A word problem that involves **equal-size groups** has to do with **multiplication**. Such a problem may ask...

- for the total number of objects, or
- for the number of groups, or
- for the number of the objects in each group.

A multiplication word problem doesn't always ask for the total! So, don't "blindly" multiply the numbers given in the problem. You need to think first. Drawing can help!

Example 1. A city has four post offices, and each post office has five workers. How many postal workers do the post offices have altogether?

This is the common situation where we have groups of equal size (groups of five workers), and there are four of them. You could draw **four** boxes (to signify the post offices), and **five** dots in each (to signify the workers).

The total number of workers is $4 \times 5 = 20$.

Example 2. Ava wrote some invitations, and put three stickers in each. She used 18 stickers. How many invitations did she write?

Here we have groups of 3 — each invitation is a "group" — but we don't know how many groups or invitations there are. We are given the total number of stickers (18), and we are asked how many groups.

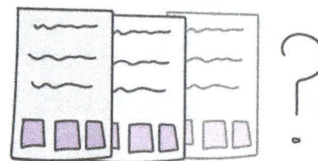

You can write a missing number multiplication sentence: $3 \times \underline{\quad} = 18.$ or the same thing using a letter for the unknown: $3 \times G = 18$. The answer is six invitations.

1. Circle the multiplication sentence that matches the problem. Then find the answer.

a. Adrian had fifteen toy cars, and he sorted them into three equal piles. How many cars were in each pile?

$15 \times 3 = \underline{\quad}$

$3 \times \underline{\quad} = 15$

b. Six children are playing tennis together. They each brought four tennis balls. How many balls do they have altogether?

$4 \times b = 6$

$6 \times 4 = b$

2. Write a multiplication sentence with an unknown for each problem. Use a letter for the unknown. You can draw pictures to help.

a. The Russ family ordered two veggie pizzas and one chicken pizza. Each pizza was sliced into four pieces. How many slices of pizza were there?

_____ × _____ = _____

There were _____ slices of pizza.

b. Each person in the Green family keeps a hand towel and a bath towel in the bathroom. In total, there are 10 towels hanging in their bathroom. How many people are in the Green family?

_____ × _____ = _____

The Green family has _____ people.

c. Mom washed laundry three times last week. In total, she used six scoops of detergent. How many scoops of detergent did she use for each wash?

_____ × _____ = _____

She used _____ scoops of detergent.

d. Jocelyn has four plastic flower pots and four ceramic ones. In each pot there are four flowers. How many flowers does she have?

_____ × _____ = _____

She has _____ flowers.

e. Sticker sheets cost $3 each. Lily buys five horse sticker sheets and two cat sticker sheets. How much did it cost?

_____ × _____ = _____

It cost $_____.

3. Solve the problems.

a. How many feet do 11 chickens have in total?	**b.** All the goats in a field have a total of 28 legs. How many goats are there?
c. If you make groups of 9 sticks, and you have 36 sticks, how many groups can you make?	**d.** Liam put eight rocks in each of his toy buckets. He had three buckets. How many rocks did he use?
e. Bridget bought a bunch of notebooks for $2 each. The total bill was $24. How many did she buy?	**f.** Tyler has saved $30. He wants to buy toy cars that cost $5 each. How many can he buy?
g. Noelia buys a bag of 40 dog treats. That day she gave each of her four dogs two treats. How many treats did she use?	**h.** An elephant walks four miles in an hour. If the elephant walks 24 miles in a day, how many hours did it spend traveling?

4. Solve the mystery numbers! (All mystery numbers are less than 100.)

Mystery Number 38 25 31 11 99 47 101 99	**a.** You can find me both in the table of eleven and five. I am _____.
b. I am more than 15. I am in the table of two, the table of three, and the table of four! I am _____.	**c.** I am in the table of 11! If you take one away from me, you will get a number that is in the table of 10. I am _____.

More Practice with Multiplication

1. Solve. Write a number sentence for each situation. Don't just write the answer.

a. Adele rides her horse for three hours, traveling 10 km each hour.

_____

She rode _____ km in total.

b. One table can seat six people, and 36 people are coming to dinner.

_____

You need _____ tables for 36 people who are coming to dinner.

c. Erasers come in packages of eight and the class has 20 children.

_____

The teacher needs _____ packages so that each child will have an eraser.

d. A set of six bath towels costs $42.

_____

You pay $_____ for each individual towel.

2. Now it's your turn! Write a word problem to fit the multiplication sentences.

a. $8 \times 4 = \triangle$

b. $3 \times \triangle = 27$

3. Solve. Write a multiplication, addition, or a subtraction sentence for each problem.

a. Naomi works in a gift shop. She packages 27 dolphin figurines into boxes and ends up with nine boxes. How many did she put in each box?	**b.** Joe bought a soccer ball for $12, but it popped after two days, so he bought a better one. He spent a total of $49 on the two. How much did the second ball cost?
c. Mariah has saved $14. She wants shoes that cost $35. How much more money does she need to buy them?	**d.** Pencils are priced at $3.00 for a box of twelve. Felix buys six boxes. How many pencils did he buy?
e. Eli has twelve pairs of socks. How many individual socks does he have?	**f.** You have $40. Umbrellas cost $14 each. How many umbrellas can you buy?
g. Brian sorted his 30 toy cars into groups of 5. How many groups did he get?	**h.** Eleven sheep each give birth to twin lambs. How many sheep are there now?

4. Solve the mystery numbers! (All mystery numbers are less than 100.)

Mystery Number 38 25 11 99 47 101 9	**a.** I am in the table of three and six. If you add two to me, I'm in the table of eight. I am _____ .
b. I am in the table of five but not in the table of ten. Adding my digits you get seven. I am _____ .	**c.** I am less than 22 but more than 9, and I am in the table of four. If you swap my digits, I am in the table of three! I am _____ .

26

Time in Word Problems

1. Solve. You can use the clocks on the side to help.

a. A rain storm lasted half an hour, starting at a quarter till 4. When did it end?

b. Music class starts at 10:15 and ends at 10:45. How long is the class?

c. It takes Mom 15 minutes to drive to the grocery store. If she leaves at 3:25, when will she arrive at the store?

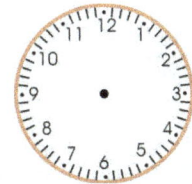

d. Dan gets to the bus stop at 9:12 AM. The next bus comes at 9:20 AM. How long does he have to wait?

e. A birthday party is starting at 6:30 PM. If it takes Joanna 15 minutes to drive there and she wants to get there 5 minutes early, at what time should she leave?

f. After a one-and-a-half hour drive, Alexa arrived at the beach at 9:30 AM. At what time did she start driving?

g. Sergio said, "Oh, I just spent 2 full hours answering emails, and now it's already 10:27!" At what time did Sergio start answering emails?

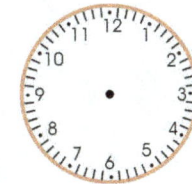

2. Today is November 6, and Ryan's birthday is in two weeks. What day is Ryan's birthday?

Mom's birthday is on December 8. How many days are between Ryan's and Mom's birthdays?

November						
Sun	Mon	Tue	Wed	Thu	Fri	Sat
1	2	3	4	5	6	7
8	9	10	11	12	13	14
15	16	17	18	19	20	21
22	23	24	25	26	27	28
29	30					

3. Today is December 12, and Michelle is going to a skiing camp for three days next Wednesday. In how many days will she go skiing?

What is the last day of the skiing camp?

December						
Sun	Mon	Tue	Wed	Thu	Fri	Sat
		1	2	3	4	5
6	7	8	9	10	11	12
13	14	15	16	17	18	19
20	21	22	23	24	25	26
27	28	29	30	31		

4. It is still three weeks to go until Harper's dentist appointment, which is on December 10. What day is it today?

5. Solve.

a. Alyssa puts an apple pie in the oven at 7:46 PM. It needs to bake for three quarters of an hour. At what time will it be done?

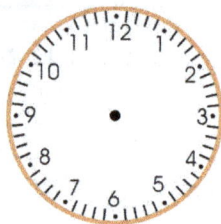

b. Melina did math from 9:05 to 9:45 AM, and later she did some extra math from 11:37 AM to 12:02 PM. For how long did she do math?

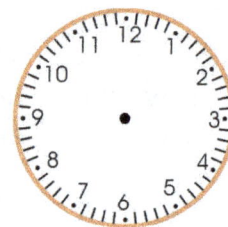

Puzzle Corner

If the sun rises at 5:24 AM and sets at 6:57 PM, for how long is it daytime?

Multiplication with Other Operations

1. How many feet do the animals have in total? Write a multiplication and addition sentence for each situation, to find the total.

a. six chickens and one cow _____ × _____ + _____ = _____	**b.** three sheep and one goose _____ × _____ + _____ = _____

c. four horses and two ducks

_____ × _____ + _____ × _____ = _____

2. Solve. Write a number sentence with several operations for each problem.

a. There were two trees with eight birds in each tree. Six of them flew away. How many birds stayed in the trees?

_____

_____ birds stayed in the trees.

b. During one week, Veronica read two books each day, except on Thursday, when she read only one book. How many books did she read in total that week?

_____

She read _____ books in total.

c. Camille bought three pairs of socks for $3 each and six pairs of cheaper socks for $2 each. How much money did she spend?

_____

She spent _____ in total.

d. A leaky sink drips two pints of water a day for eight days. Then the leak gets worse, and it drips *three* pints of water a day for 11 more days before Dad fixes it. How many pints of water were wasted?

_____

There were _____ pints of water wasted.

3. Solve. Write a number sentence with several operations for each problem.

a. The Gordons ordered four pizzas, sliced into four pieces each.
The dog ate one piece. How many pieces did the people get?

b. Jenna plants six rows of daffodil flowers with twelve in each row.
Of all the seeds, 63 daffodils sprout. How many seeds didn't sprout?

c. Mateo earns $7 every time he helps with the yard work. He did yard work
seven times and saved all his money. Then he bought a book that cost $15.
How much money did he have left?

4. Solve.

a. If a puffin catches about 10 fish in a trip, and eats 40 fish a day, how many
fishing trips would it make in a week?

b. You see chickens and cats walking on the yard, and they have a total of 22 legs.
How many cats and how many chickens are there?

(*Challenge*) Find two other solutions to the the problem.

Puzzle Corner Hannah has 62 chickens. She goes through one 50-pound sack
of chicken food in four days. How many sacks of food will she
use in four weeks?

Mixed Word Problems 1

1. Solve. Write a number sentence to go with each problem.

a. In a garden, there are eight ladybugs that each have seven spots and two ladybugs that have nine spots. How many spots do the ladybugs have in total?

b. Teddy the dog has the zoomies. He runs one circle around his owner every 3 seconds. How many times will he circle his owner in 30 seconds?

Now Teddy slows down and makes one circle every five seconds. How many times will he circle his owner in the next 30 seconds?

c. A beaver building a dam cuts down a tree every nine minutes. If it starts cutting trees at 8:15 AM, how many will it have cut down by 9:00 AM?

d. If a hummingbird drinks nectar from a flower once every fifteen minutes during the day, how many times will it feed over the course of six hours?

e. Michelle is baking blueberry muffins for a party. The muffins take 25 minutes to bake in the oven, but she can't fit all of them in the oven at once, so she decides to bake them in two batches. If she starts baking them at 2:30 PM, when will both batches be done?

2. Gary owns a horse farm. The graph shows the number of horses he has of each color.

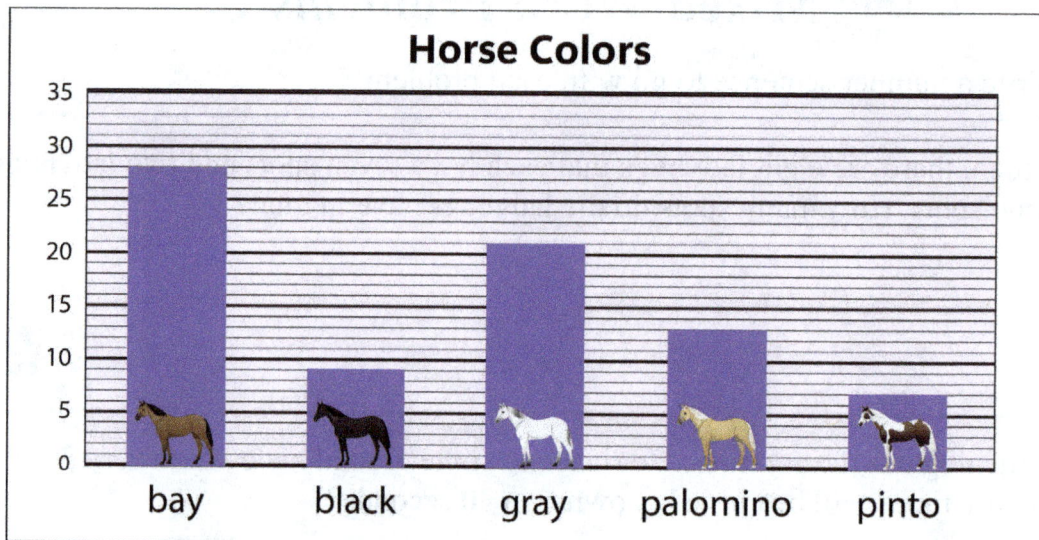

Horse Colors

a. Which color does he own the fewest of?

b. How many more bay horses are there than palominos?

c. Gary sells two gray horses and a black horse.
How many horses does he have in total now?

3. Solve.

Tim bought a $21 turtle pool float. Then he left it in the car and forgot about it, so a week later, he bought a $26 dolphin pool float.

On the right, you see all the bills he used to pay for these two things. What was the change he received in total?

First purchase:

Second purchase:

Money and Change

Change is the **difference** between the total cost of the items and the money given by the customer. There are two basic ways to calculate change:

(1) Count up from the total cost until you reach the amount given by the customer.

(2) Subtract the amount given by the customer and the total cost.

Example. Arnold bought a $46.75 jacket and paid with $100. What was his change?

(1) Count up:	**(2) Subtract:**
From $46.75 to $47: **25 cents** From $47 to $50: **$3** From $50 to $100: **$50** **Total: $53.25**	$$\begin{array}{r} {\scriptstyle 9\ \ 9\ \ \ 9} \\ {\scriptstyle \cancel{10}\ \cancel{10}\ \ \cancel{10}\ 10} \\ \$\cancel{1}\ \cancel{0}\ \cancel{0}\ .\ \cancel{0}\ \cancel{0} \\ -\ \ \ 4\ 6\ .\ 7\ 5 \\ \hline \$\ 5\ 3\ .\ 2\ 5 \end{array}$$

Since subtracting to find change usually involves regrouping over many zeros, it is error-prone. Adding up is often easier.

1. Solve.

a. Can you buy two boxes of crayons for $2.40 each with $5?

If not, find how much more you would need.

If yes, find your change if you buy them.

b. Can you buy a $3.80 energy bar and a $1.50 apple juice with $5?

If not, find how much more you would need.

If yes, find your change if you buy them.

c. Can you buy three pineapples costing $3.70 each with $10?

If not, find how much more you would need.

If yes, find your change if you buy them.

2. Calculate.

a. Cole opened his piggy bank and counted the coins. He had 245 cents. He also had $5 in his wallet. How much money does Cole have in total?	**b.** Cadence buys a mini animal eraser for 35 cents, cat stickers for 99 cents, and a pen for $2.05. How much is it in total?
c. Evangeline didn't have any small change, so she paid for her lunch with $50, and got $31.50 in change. How much did her meal cost?	**d.** Alex bought three slices of pizza for $4.35 each. What was his change from $15?

e. Holly bought yogurt. She paid with $10 and got $5.87 in change. Was it the correct amount of change? If not, correct it.

$5.23

f. Talon paid for his soup with $15 and the clerk gave him three dollars, four dimes, and two pennies. Did he get the right amount of change? If not, correct it.

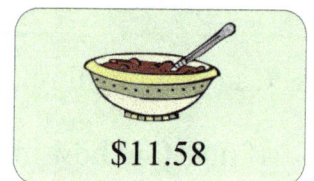

$11.58

34

3. Solve the problems.

| $8.50 | $4.85 | $3.15 | $11.58 | $5.23 |

a. Suzy bought a sandwich and a pizza slice. Find the total price and her change from $8.

b. Hungry Hector bought a sandwich, soup, and yogurt. He paid with $30. What was his change?

c. The Green family loves to eat salad. All four family members ordered salad for lunch. How much did it cost?

d. Which is cheaper: salad and yogurt, or soup and pizza?

More Multiplication

1. Solve. Write a number sentence for each problem.

a. Elijah bought six notebooks for $3 each and a package of pencils for $4.
What was the total cost?

_____ × _____ + _____ = _____

b. Each minibus holds twelve passengers. There are five full minibuses, and two
with one empty seat. How many passengers are there in total?

c. You can fit seven dominoes in one layer in their box. How many layers will
there be with 28 dominoes?

d. Mom uses lots of eggs. She buys three dozen eggs, but she also has four eggs left
at home. How many eggs did she buy?

e. Aiden bought some packages of seeds for $18. Each package cost $2.
How many packages did he buy?

f. Roses are sold in bunches of five. Dad bought six bunches
and four extra roses for Mom's birthday—a rose for each year.
How old is Mom?

2. Solve.

a. In one part of a reef, there are four groups with six clownfish each, five groups with five clownfish, and two groups of two clownfish. How many clownfish are there in total?

b. Miguel had $6 in his piggy bank. Each week, for seven weeks, he saved $3 from the money he earned. Afterwards, he had just enough money to buy a model airplane. How much did the airplane cost?

c. A can of three tennis balls costs $4. How many cans can you buy if you have $30 but want to have at least $10 left over afterwards?

Puzzle Corner

Two frogs, Mack and Louie, decide to race across the small pond they live in, which is 42 feet wide.

- Louie crosses the pond by jumping across lily pads on top. He covers seven feet in each jump and each jump takes one second, but he has to stop and rest for three seconds in between each jump.

- Mack swims across the pond with a speed of six feet every four seconds.

How many seconds will Mack take to cross the pond?

How many seconds will Louie take to cross the pond?

Which frog won the race?

37

Mixed Word Problems 2

1. Solve.

a. Grandma's house is 30 miles away. After driving halfway there, Mom realizes she left the oven on and drives back home. Then she drives to Grandma's, stays for 3 hours, and goes back home. How many miles did Mom drive?

b. Levi has a toy marble run. He puts 11 marbles on the marble run four seconds apart, and each marble takes one minute to finish all the slides. How many minutes and seconds will it take for all the marbles to finish?

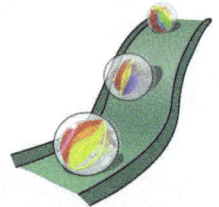

c. Farmer Steven harvests 444 tomatoes. 396 of them are good quality, and the rest are odd-shaped or have brown spots. He plans to sell the latter as second-grade tomatoes in bags of eight each, costing $1.50 a bag.

How many bags will he get?

If he sells all the bags, how much money will he make from the second-grade tomatoes?

2. Make a QUESTION for each situation. (Think what you can find out using what the problem tells you.) Then solve your question.

a. Brooklyn wants to read four books that have 340, 196, 153, and 216 pages.	**b.** Grant tied balloons together in bunches. He made seven bunches of four. The eighth group had twelve balloons.

c. For two weeks, Fred the cat catches two mice a day. Then the mice get smart and become harder to catch, so on the third week Fred only catches one mouse a day.

d. A desk costs $154 and some chairs cost $128.

3. Solve the mystery numbers! (All mystery numbers are less than 100.)

Mystery Number 38 25 11 99 47 101	**a.** I am between 49 and 61. If you subtract the sum of my digits from me, the new number will have the same sum of digits. I am _____.
b. I am between 15 and 85. If you add my digits, you get the same answer as if you multiply my digits. I am _____.	**c.** If you first double me, and then multiply my digits, you will get 32. I am in the table of seven. What am I? I am _____.

Word Problems with Large Numbers

1. Solve.

a. A flock of 80 geese migrate 2,380 miles south
one winter, and then fly back in the spring.
How many miles did they fly during those trips?

b. Cameron bought two cameras for a total of $1,929.
The cheaper camera cost $500.
How much did the other camera cost?

c. A farmer gets 4,200 eggs one year from his chickens.
The next year, he gets seven more chickens and gets
5,600 eggs. If he gets seven more chickens again the
year after that, how many eggs can he expect to get?

d. Weston sends a package in the mail to his grandparents.
The package travels 6,200 km in an airplane across the
ocean, then 1,400 km in another airplane, and then 140 km
by car before being dropped off at Grandma's house.
How far did the package travel?

2. Solve.

a. A jogging track is 3,500 feet, but there is a shortcut that shortens it by 840 feet. Tristan jogs around the track using the shortcut twice. How far did he jog in total?

b. Three different saddles cost $1,200, $1,990, and $3,550. What is the price difference between the cheapest and the most expensive saddle?

c. One piano costs $2,890, another $3,640, and a third one $4,150. A principal bought two of the cheapest pianos and the most expensive one for his school. What was the total cost?

d. Ashley is saving up to buy a car that costs $7,590. She wants to have at least $1,500 left over after buying it. Right now she has $6,324. How much more does she need?

Geometry Puzzles

1. Solve.

a. A grocery store is 415 feet long and 144 feet wide. What is its perimeter?

b. A big red rug covers all of a room that measures 9 × 12 feet. What is its area?

2. Solve. In (a), you can draw the situation to help.

a. Axel tiled a room using 32 tiles. Each tile had an area of two square feet. What was the area of the room?

b. One side of a room is 6 meters, and the perimeter is 20 meters. What is the area?
(Hint: First find the length of the other side of the room.)

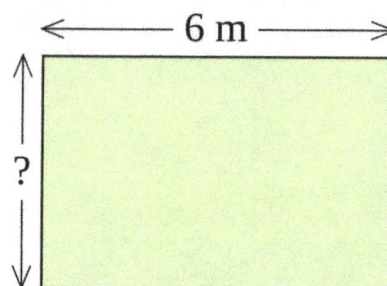

← 6 m →

?

c. A shed has an area of 42 square feet. One side is seven feet long. What is the perimeter?

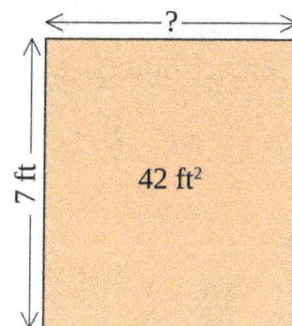

← ? →

7 ft   42 ft²

3. Solve.

a. An 8×10-inch painting is in a frame with a width of 10 inches and height of 12 inches.
What is the area of the painting (the yellow area)?

What is the area of ONLY the frame (the gray area)?

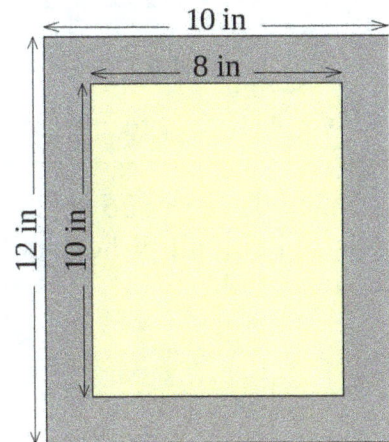

b. Milo wants chickens. He wants 19 hens and one rooster. He wants to have at least four square meters of outdoors space per chicken, plus a 3×3-meter coop. What is the minimum amount of space he needs for them, in square meters?

Milo is still farming and wants a rectangular garden. He starts by building a fence to keep his chickens out of the vegetables he plans to grow. He has 20 meters of fencing he can use for the perimeter. How long should each side be to give him the biggest possible area of garden?

Puzzle Corner

Word Problems with Measurement Units

To the teacher: use these worksheets only if your student(s) have studied measurement unit conversions.

1. Solve. Ask your teacher if you need a reminder on measurement unit conversions.

a. Makayla is 1 m 35 cm tall and Savannah is 134 cm tall. Who is taller, and by how much?	**b.** A square window has sides that are 3 feet 8 inches each. What is the perimeter of the window?
c. A light bulb weighs 50 grams. How many of those would weigh one kilogram?	**d.** A baseball weighs 5 oz. How much would six baseballs weigh?
e. Colette uses 20 ml of shampoo a week on average. She buys a 1-liter shampoo bottle. How many milliliters of shampoo will be left in eight weeks?	**f.** Three different parrots weigh 880 grams, 650 grams, and 1 kg 100 g. What is their total weight put together, in grams?
g. Ethan lines up his three big toy cars, each 11 inches long, and four smaller cars, each four inches long. How many feet and inches long is the line?	**h.** Two young trees are two meters tall. Four years later, one of the trees has grown a meter and the other grew 28 cm more than that. How many centimeters tall is the taller of the two trees now?

2. Solve.

a. A group of different kinds of chocolate bars weighs a total of 3 pounds. Half of this is in bars that weigh 2 oz each, and the rest of the bars weigh 3 oz each. How many 3-oz chocolate bars are there?

b. Roger's family bought a 2-liter jug of orange juice. The next day, there is 1 L 100 ml left. Then Roger drinks another 250 ml. How many milliliters of orange juice has his family drunk so far?

c. A tortoise is walking 23 yards to eat cucumbers in Milo's garden. It has walked 35 feet so far. How many more feet does it have to go?

Dad buys a half-gallon tub of ice cream. If Dad eats one cup of ice cream once every two days, and Ella and Alana each eat half a cup of ice cream once every two days, how many days will the ice cream last?

(Reminder: one gallon is four quarts, one quart is two pints, and one pint is two cups.)

45

Division in Word Problems

When deciding between using division or multiplication for a word problem, check what the problem is asking. Is the problem asking for a total? Or do you already know the total, and it asks "how many groups/parts" or "how many in each group/part"?

It can also help to think about the problem as a multiplication sentence first.

Example. Zara drove to a zoo and back, a total of 60 miles. How far away is the zoo?

You already know the total amount of miles (60). So, you can set up a multiplication sentence that reads **2 × _____ = 60**. Then you can reverse the problem to be a division sentence: $60 \div 2 = 30$.

1. Write a multiplication and a division sentence for each situation.

a. Cecilia put 16 toys into boxes, four toys in each.	**b.** Sadie planted eight rows of six daffodils each.	**c.** A total of 108 apples were packaged into nine bags.
_____ × _____ = _____	_____ × _____ = _____	_____ × _____ = _____
_____ ÷ _____ = _____	_____ ÷ _____ = _____	_____ ÷ _____ = _____

2. Write a multiplication OR a division sentence for each problem.

a. Keira buys ten apples. She gives five apple slices to each of twelve horses. How many apple slices were there total?

b. There are 66 angelfish in one part of a coral reef living in 11 different groups. How many are in each group?

c. Amanda, Emma, and Isla shared eighteen marbles evenly. How many did they each have?

3. Solve the problems. Write a number sentence or several for each.

a. A set of three mugs costs $18. How much does each mug in the set cost?

b. Kelly packaged 64 cookies evenly between eight boxes and then sold six of the boxes. How many cookies does she have left?

c. Hunter buys $27 worth of energy bars for $3 each. Then he eats some, and has six bars left. How many did he eat?

d. Five kids shared evenly three small pizzas, each cut into six slices. The dog got the slices they couldn't share evenly. How many slices did the dog get?

e. Mackenzie is filling a 180-gallon water trough with a hose that supplies nine gallons every minute. How long will it take to fill?

f. Bill, Dill, Will, and Phil shared 30 marbles unevenly so that Bill and Dill each have three more marbles than Will and Phil. How many does each person have?

Mixed Word Problems 3

1. Solve.

a. A tortoise got into Milo's garden at 4:30 PM and is eating one ounce of cucumber every three minutes. Milo is coming home from work at 5:30 PM to water his garden. How many ounces of cucumber will the tortoise have eaten by then?

If each cucumber weighs 13 oz, how many whole cucumbers will the tortoise have eaten?

b. One morning, a rooster had two crowing sessions. In the first session, he crowed at 12:02 AM, 12:03 AM, 12:06 AM, 12:13 AM, 12:14 AM, and 12:15 AM.

In the second session, he crowed at 4:43 AM, 4:44 AM, 4:45 AM, 4:53 AM, and 4:59 AM.

Which of his crowing sessions was longer?

How long did he spend without crowing between his two crowing sessions?

c. Two months ago, Pudding the parrot knew twice as many words as Sushi. Sushi knew 12 words.

Now, Pudding has doubled how many words he knows. Sushi has learned one more word than Pudding learned in these two months.

How many words did Pudding know two months ago?

How many words does Pudding know now? Sushi?

2. A horse barn is 9 × 15 meters and has six stalls, each 3 × 4 meters.
   What is the area of the barn that is NOT stalls (the yellow area)?

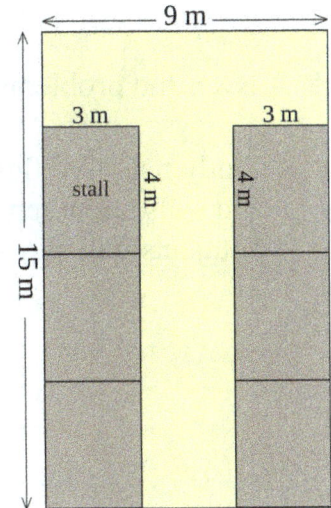

3. Make your own word problems to match the given number sentences!
   Then, give the problems to a friend or family member to solve.

a. 3 × 12 + 20

b. 50 − 3 × 8

Division and Other Operations

1. Answer the problems using multiplication or division. Use mental math.

a. Cindy puts five flowers in each bouquet and sells about ten a day. How many bouquets will she get out of 55 flowers?	**b.** Milo's chickens laid 39 eggs this week. He put them into cartons by the dozen. How many cartons did he get? How many eggs didn't fit into cartons?
c. If four people fit in a taxi, how many people would fit in nine taxis?	**d.** Nadia fills eight pages of her sticker book with 64 stickers. How many stickers are on each page?

e. A really big apple crisp recipe calls for 3 cups of oats and for 15 apples. You're making a smaller batch, so you only use 1 cup of oats. How many apples do you use then?

2. Make your own word problem using the given situation. You will need to continue the story and add more information to each situation.

a. There are 18 bones and some dogs.

b. Milo has grown 67 cucumbers. He eats some himself, feeds some to his six chickens, and gives the rest to his neighbors.

3. Solve the problems.

a. A panda eats 21 pounds of bamboo in 7 hours. How much does it eat in five hours?

Hint: First answer the question: how much does it eat in one hour?

b. Twelve speedboats are carrying a total of 63 people. Three of them have six people each. The rest carry the same number of people each. How many do each of the others have?

c. Morris wants lots of canned fish. He buys $80 dollars worth of tuna cans costing $2 each and $60 dollars worth of sardine cans costing $3 each. How many cans did he buy in total?

Matthew earns $12 every time he mows the lawn. He also earns $40 a month for other yard work. After four months, he buys a watch that costs $125. Now he has $95.

How many times did he mow the lawn?

Fraction Problems

1. Solve. Drawing can help!

a. Danny ate 1/4 of a pie and Johnny ate 1/3 of it. Who ate more?

b. Dad eats 2/10 of a bowl of blueberries and Mom eats 7/8 of a bowl. Who ate more?

c. Valerie plants 15 flowers, of which 7 are bluebells, 5 are daffodils, and the rest are dahlias. What fraction of them are dahlias?

Miranda also has 15 flowers, and 1/3 of them are daffodils. Does she or Valerie have more daffodils?

d. Frank and Max have two same-sized pizzas. Frank eats 2/5 of his. Max cuts his pizza into ten slices. How many slices should he eat to eat the same amount as Frank?

e. Gerald, Hank, Xavier, and Ricardo each have chocolate bars of the same size. Gerald eats 2/12 of his chocolate bar, Hank eats 1/5 of his, Xavier eats 1/6, and Ricardo eats 2/10. Who ate the same amount as Gerald?

2. Solve. Drawings can help!

a. You have a BIG plateful of chicken salad, and your sister has a little plateful. You both eat 3/4 of your plate and get full. Who ate more salad?

b. A panful of cornbread was cut into 12 equal pieces. Julie ate 3 of the pieces, which was 3/12 of the cornbread. Which other, easy fraction is equal to 3/12?

c. Mom cut a bunch of apples into sixths. Abby ate three entire apples plus one sixth, and Susanna ate 17 sixths. Who ate more?
How much more?

d. A bunch of tomatoes were cut into fourths. Andy ate nine fourths, and his brother Barry ate two entire tomatoes plus 3/4 of a tomato.
Who ate more?
How much more?

e. Charlotte bought 12 notebooks. Half of them were blue, 1/6 of them were yellow, and 1/3 were green.

How many were green?

How many were blue?

How many were yellow?

f. Who ate more? In which one of these situations can you NOT easily tell who got more?

(i) Eddy ate 1/4 of this pie , and Chelsea ate 1/6 of this pie: .

(ii) Greg ate 3/8 of this pizza: , and Vincent ate 3/5 of this pizza: .

Word Problems with Large Numbers 2

1. Solve.

a. While flying, a falcon ascends 238 feet and then descends 412 feet. Now it is flying at a height of 2,852 feet. How high was it flying before?

b. Two different horses cost $1,955 and $9,590. What is the price difference? Estimate first:

Estimation: _____

Exact answer: _____

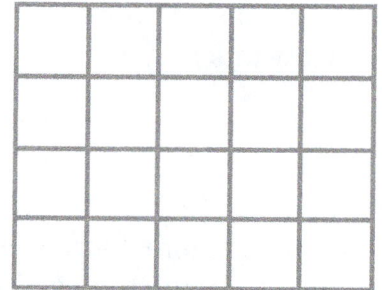

c. An adult ticket for a certain flight is $637 and a child's ticket is $395. Find the total cost of two adult tickets and three children's tickets. Estimate first:

Estimation: _____

Exact answer: _____

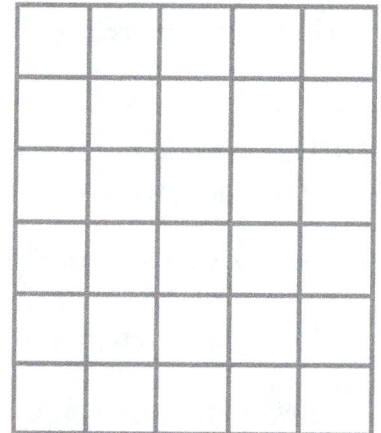

d. Willow bought fence posts for $1,784 and fence rails for $1,697. Then she got a $499 discount. She paid with $3,000. What was her change?

2. Below are the approximate weights of various animals, in pounds.

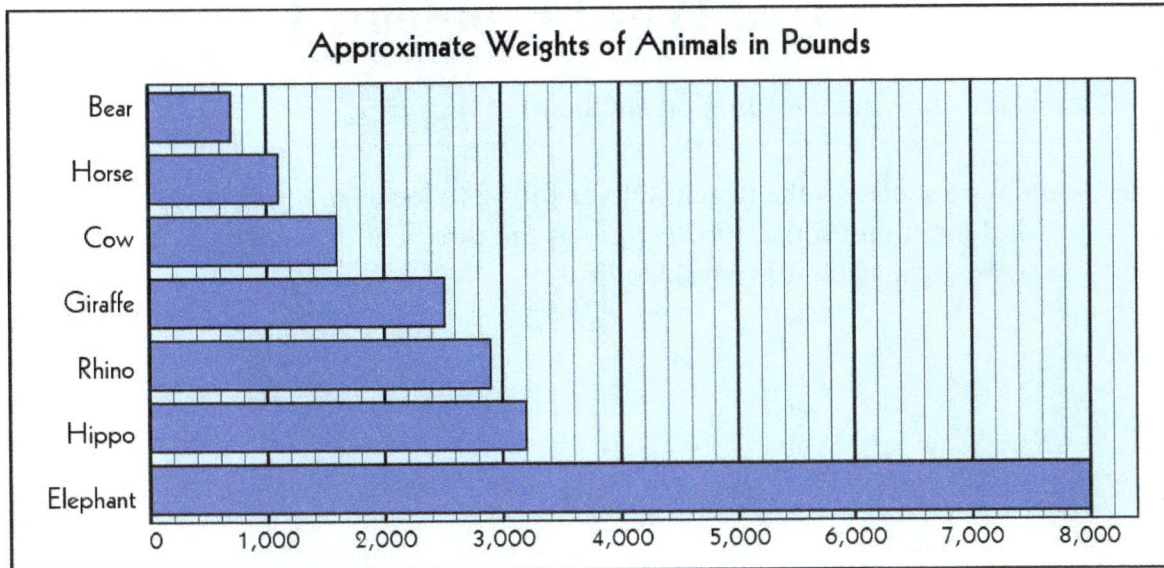

Approximate Weights of Animals in Pounds

Animal	Weight
Bear	~700
Horse	~1,500
Cow	~2,000
Giraffe	~2,500
Rhino	~3,000
Hippo	~3,500
Elephant	~8,000

(Bar graph with x-axis scale: 0, 1,000, 2,000, 3,000, 4,000, 5,000, 6,000, 7,000, 8,000)

a. How much heavier is the giraffe than the horse?

b. How much heavier is the elephant than the giraffe?

c. How many hippos would you need to outweigh the elephant?

d. Which two kinds of animals have the same combined weight as two other kinds of animals?

Puzzle Corner

On a small island, there are 4,241 penguin chicks, plus adult penguins. If each *pair* of adult penguins had two eggs, and all but 231 of those eggs hatched, how many penguins are there?

Hint: You know how many chicks there are, so first find out how many adults there are.

55

Mixed Word Problems 4

1. Solve. Remember, drawing the situation can help!

a. Asher found nine shells on the beach. Oliver found twice as many as Asher. Marlene found some, too. Together, the three children found 44 shells. How many did Marlene find?

b. One set of glitter pens costs $11 and another costs $3 more. Shirley buys three of the cheaper sets and pays with $50. Find the total cost.

c. A box of cookies says that one serving of three cookies has 15 grams of sugar. Clara wants to eat less than 20 grams of sugar a day, and she has already eaten 8 grams of sugar. How many whole cookies can she eat without going over 20 grams?

d. For a week, Marvin alternates reading two short stories a day and four a day, starting with reading two stories on the first day. How many has he read by the end of the week?

2. Benny the dog has a sack of 12 bones. He wants them to last as long as possible, and decides to dig up the yard and hide the bones at equal distances along all four sides of the fence line, where the dirt is the softest. If the sides of the yard measure 24 ft and 30 ft, how far apart will the bones be?

Hint: Draw a sketch of the yard.

3. Make your own word problem about strawberries! Make it have more than one step (more than one operation). You can give it to a friend or family member to solve.

Puzzle Corner Luca bought a toy truck for $7.83 and paid with $10. However, the clerk was low on change and didn't have any pennies, so she gave him $2.20 in change. Since this was too much change, he gave her some more money. How much more did he pay the clerk?

Answer Key

Page 7

Example. Eli paid $40 LESS than $650. It means we SUBTRACT: $650 − $40 = $610. The final price was $610.

1. a. 15 + _?_ = 62. Answer: 47 glitter pens work.

   b. 148 + 4 = _?_. Answer: Cash traveled 152 km.

   c. 74 − _?_ = 46. Answer: There is a 28-degree difference in temperature.

   d. 54 − 44 = _?_. Answer: Zoe needs 10 more pins.

Page 8

2. a. 46 + ___ = 143 or 143 − 46 = 97. 97 seats are not window seats.

   b. 175 − 4 − 9 = 162 or 4 + 9 = 13 and 175 − 13 = 162. 162 people did come to the meeting.

   c. 365 − 104 = ___ or 104 + ___ = 365. It did not rain on 261 days.

   d. 14 + 4 = 18 , and 25 − 18 = ___ or 18 + ___ = 25. Elliott has 7 more cars than Owen.

   e. 74 + ___ = 80 or 80 − 74 = ___. Blake still needs $6.

Word Problems: Mental Math, pp. 9-10

Page 9

1. a. 11 + 6 + 3 + 7 = 27 or x − 11 − 6 − 3 = 7. There were 27 cookies originally.

   b. 12 − 5 + 10 = 17. There are now 17 children on the playground.

   c. The boys have 24 + 16 = 40 marbles together. Yes, they can share those evenly (since 40 is an even number). Each boy would get 20 marbles.

Page 10

2. a. 131 − 4 = x., and 131 − x = 4. Molly is 127 cm tall.

   b. 133 + 6 = B, and B − 6 = 133. Bentley is 139 cm tall now.

3. a. 50 + 15 + 30 = x. The whale swam 95 miles.

   b. 40 − 8 − 8 = 24, OR 8 + 8 = 16 and 40 − 16 = 24. You have $24 left.

4. a. 35 + 5 + 8 = x. Holly now has $48.

   b. 20 − x = 9, or 20 − 9 = x. The gift cost $11.

   c. 199 + 12 = x. The price is now $211.

   d. 250 − 211 = x, or 211 + x = 250. Her change was $39.

Multiple Steps — Addition and Subtraction, pp. 11-13

Page 11

1. $25 - 8 = 17$, so the cheaper watch costs $17. Then add $17 + 17 = 34$. The total cost is $34.

Page 12

2. a. $64 - 9 = 55$, so Dante now has 55 balls. $55 - 31 = 24$. <u>Dante has 24 more golf balls than Jace.</u>

   b. $320 + 20 = 340$, so Brady needs to save up to $340. $340 - 270 = 70$. He needs to save $70 more.

   c. $11 + 23 + 14 = 48$, so Nadia needs to save up to $48. $48 - 19 = 29$. Nadia has $29 right now.

   d. $167 + 185 = 352$. There will be a total of 352 squares. $352 - 138 = 214$. She has 214 squares left to sew.

Page 13

3. a. $25 + 25 + 30 + 40 - 87 = 33$. Mr. Gale has 33 kg of fruit left.

   b. $115 + 8 = 123$. It costs $123 to fix a leaky sink this year. $123 + 123 + 123 = 369$. It costs the Howards $369.

   c. Elk Valley School has $727 - 116 = 611$ students. Lakeside school has $338 + 116 = 454$ students. <u>Elk Valley School has more students.</u> They have $611 - 454 = $ <u>157 more students.</u>

More Addition and Subtraction Word Problems, pp. 14-15

Page 14

1. a. $56 - 27 - 15 = x$, or $27 + 15 = 42$ and $56 - 42 = x$. Kara needs $14 more.

   b. $64 + 64 + 64 - 13 = 179$. There are 179 pencils in the boxes.

   c. $429 - 30 = 399$. The second laptop costs $399. $429 + 399 = 828$. The laptops cost $828.

   d. $95 + 95 + 95 - 130 = 155$. Ken has $155 left.

Page 15

3. a. $355 + 355 + 52 = 762$. The trees grow 762 limes.

   b. $400 - 115 - 75 = 210$. There are 210 blue sweaters.

4. Answers will vary. Check the student's answer.

   a. EXAMPLE ANSWER: There are 14 birds in a tree. A few minutes later, some more come, and now there are 31 in the tree. How many just flew in? Answer: 17.

   b. EXAMPLE ANSWER: Eli spends $32 on two soccer balls and now has $105 left. How much did he have before? Answer: $137.

Extra Numbers in Word Problems, pp. 16-17

Page 16

1. a. $332 + x = 489$. $x = 489 - 332 = 157$. Theo needs $157 more.

   b. $43 + 286 + 121 = 450$. The calf weighs 450 kg now.

   c. $369 + 369 = 738$. The total cost is $738.

   d. $25 + 13 + 18 = 56$. Finley bought 56 birds.

Page 17

3. a. $785 + 137 = 922$. She earned $922.

   b. $164 + 164 + 56 = 384$. There are 384 pieces in the pile.

   c. Leo chewed up $41 - 8 = 33$ bones the second month and $3 + 2 = 5$ shoes the second month. $41 + 33 + 3 + 5 = 82$. Leo chewed up 82 bones and shoes in total.

   d. The farmer has $92 + 3 + 163 = 258$ sheep before selling any. $258 - 8 - 139 = 111$. Now he has 111 sheep.

Rounding, pp. 18-19

Student estimates may vary. The estimates given here are just one possibility. However, it does make most sense to round three-digit numbers to the nearest ten.

Page 18

1. a. Estimation: $140 + 20 = 160$ or $140 + 16 = 156$. Answer: $139 + 16 = \underline{155}$.
   b. Estimation: $\$40 + \$30 + \$30 = \$100$. Answer: $\$38 + \$29 + \$29 = \underline{\$96}$.
   c. Estimation: $\$780 - \$150 = \$630$. Answer: $\$776 - \$147 = \underline{\$629}$.

Page 19

2. Estimate: 240 ft + 280 ft + 130 ft = 650 ft. Answer: 243 ft + 127 ft + 281 ft = $\underline{651 \text{ ft}}$.

3. a. About 230 km + 230 km = $\underline{460 \text{ km}}$. (The exact answer: 466 km)
   b. About $\$130 + \$120 + \$210 = \underline{\$460}$. (The exact answer: $454)
   c. About $370 - 180 = \underline{190 \text{ days}}$. (The exact answer: 187 days)
   d. About $360 - 290 = \underline{70 \text{ pages}}$. (The exact answer: 72)
   e. The two phones that cost $369 and $512 cost about $\$370 + \$510 = \$880$ together (actually $881).
      The two phones that cost $445 each are about $\$450 + \$450 = \$900$ together (actually $890).
      The first deal is better.

Money Situations, pp. 20-21

Page 20

1. a. $34 + 34 = 68$, and $34 + 34 + 34 = 102$. You can buy two backpacks with $100. $100 - 68 = 32$. You will have $\underline{\$32}$ left.
   b. $8 + 8 = 16$, and $8 + 8 + 8 = 24$, and $8 + 8 + 8 + 8 = 32$, and $8 + 8 + 8 + 8 + 8 = 40$. Or, $5 \times 8 = 40$.
      You can buy five mugs with $40. You will have $\underline{\$0}$ left.
   c. He has $10, so needs to earn $159 more. $40 + 40 = 80$, and $40 + 40 + 40 = 120$, and $40 + 40 + 40 + 40 = 160$.
      He will have enough money to buy the tablet in underline four weeks.

Page 21

2. Student estimations may vary.
   a. Estimated total: $\$17 + \$40 = \$57$. Exact total: $54.70
   b. Estimated total: $\$155 + 40 = \$195$. Exact total: $193.44

3. a. $2.40 + 2.40 + 2.40 + 2.40 = 9.60$. Ernest can buy four bottles of glue.
   b. $\$28.50 + \$21.50 = \$50$. Yes, the change was correct.
   c. The microscope is about $40 and the mice are about $8 a piece. $8 + 8 + 8 + 40 = 64$. Yes, Carol can buy them with $70.

Multiplication in Word Problems, pp. 22-24

Page 22

1. a. $3 \times \underline{\hspace{1cm}} = 15$. There are $\underline{5}$ cars in each pile.
   b. $6 \times 4 = b$. There are $\underline{24}$ balls.

Page 23

2. a. $\underline{3 \times 4 = y}$. There were $\underline{12}$ slices of pizza.
   b. $\underline{2 \times z = 10}$. The Green family has $\underline{5}$ people.
   c. $\underline{3 \times s = 6}$. She used $\underline{2}$ scoops of detergent for each wash.
   d. $\underline{4 \times 8 = f}$. She has $\underline{32}$ flowers.
   e. $\underline{3 \times 7 = s}$. It cost $\underline{\$21}$.

Page 24

3. The student's number sentences may or may not include an unknown.

    a. $2 \times 11 = 22$. There are <u>22</u> chicken feet.

    b. $4 \times 7 = 28$. There are <u>7</u> goats.

    c. $9 \times 4 = 36$. You can make <u>4</u> groups.

    d. $3 \times 8 = 24$. He used <u>24</u> rocks.

    e. $2 \times 12 = 24$. She bought <u>12</u> notebooks.

    f. $5 \times 6 = 30$. He can buy <u>6</u> toy cars.

    g. $2 \times 4 = 8$. She used <u>8</u> treats.

    h. $4 \times 6 = 24$. The elephant spent <u>6</u> hours traveling.

4. a. 55

    b. 24 is the intended answer, since the children have likely studied the table of 2 only up to 24. However, in reality the following numbers also fit the description: 36, 48, 60, 72, 84, and 96.

    c. 11

More Practice with Multiplication, pp. 25-26

Page 25

1. The student's number sentences may or may not include an unknown.

    a. $3 \times 10 = 30$. She rode <u>30 km</u> in total.

    b. $6 \times 6 = 36$. You need <u>6</u> tables for 36 people who are coming to dinner.

    c. $2 \times 8 = 16$, $3 \times 8 = 24$. The teacher needs <u>3</u> packages so that each child will have an eraser.

    d. $6 \times 7 = 42$. You pay <u>$7</u> for each individual towel.

2. Student answers will vary.

    a. EXAMPLE ANSWER: There are eight horses in a paddock. How many legs are there? Answer: 32.

    b. EXAMPLE ANSWER: Andrew buys three shirts for $27. How much was each shirt? Answer: $9.

Page 26

3. The student's number sentences may or may not include an unknown.

    a. $9 \times 3 = 27$. She put <u>3</u> in each box.

    b. $49 - 12 = 37$. The second ball cost <u>$37</u>.

    c. $14 + 21 = 35$. She needs <u>$21</u> more.

    d. $6 \times 12 = 72$. He bought <u>72</u> pencils.

    e. $2 \times 12 = 24$. He has <u>24</u> socks.

    f. $14 + 14 = 28$; $14 + 14 + 14 = 42$. You can buy <u>2</u> umbrellas with $40.

    g. $5 \times 6 = 30$. He got <u>6</u> groups.

    h. $3 \times 11 = 33$, or $2 \times 11 + 11 = 33$. There are <u>33</u> sheep now.

4. a. 30   b. 25   c. 12

Time in Word Problems, pp. 27-28

Page 27

1. a. 4:15   b. 30 minutes   c. 3:40   d. 8 minutes   e. 6:10   f. 8:00   g. 8:27

Page 28

2. Ryan's birthday is on <u>November 20</u>. There are <u>20 days</u> between Ryan's and Mom's birthdays.

3. She will go skiing in <u>four days</u>. The last day of skiing camp is on <u>December 18</u>.

4. It is <u>November 19</u>.

5. a. 8:31 PM   b. 65 minutes

Puzzle corner: 13 hours and 33 min

Multiplication with Other Operations, pp. 29-30

Page 29

1. a. $6 \times 2 + 4 = 16$ b. $3 \times 4 + 2 = 14$
   c. $4 \times 4 + 2 \times 2 = 20$

2. a. $2 \times 8 - 6 = 10$. <u>10</u> birds stayed in the trees.
   b. $2 \times 7 - 1 = 13$, OR $2 \times 6 + 1 = 13$. She read <u>13</u> books in total.
   c. $3 \times 3 + 6 \times 2 = 21$. She spent <u>$21</u> in total.
   d. $2 \times 8 + 3 \times 11 = 49$. There were <u>49</u> pints of water wasted.

Page 30

3. a. $4 \times 4 - 1 = 15$. The people got <u>15</u> slices.
   b. $6 \times 12 - 63 = 9$. There were <u>9</u> seeds that didn't sprout.
   c. $7 \times 7 - 15 = 34$. He has <u>$34</u> left.

4. a. $10 \times 4 = 40$, so the puffin makes 4 trips a day. $4 \times 7 = 28$. The puffin makes <u>28</u> trips in a week.
   b. Possible solutions are: 2 cats and 7 chickens, 3 cats and 5 chickens, and 4 cats and 3 chickens.

Puzzle corner: There are $4 \times 7 = $ <u>28</u> days in four weeks. $4 \times a = 28$. She goes through <u>7</u> sacks of food in four weeks.

Mixed Word Problems 1, pp. 31-32

Page 31

1. The student's number sentences may or may not include an unknown.
   a. $8 \times 7 + 2 \times 9 = 74$. The ladybugs have <u>74</u> spots.
   b. $3 \times a = 30$. He will circle his owner <u>10</u> times in 30 seconds.
      $5 \times b = 30$. Now he will circle his owner <u>6</u> times every 30 seconds.
   c. $9 \times t = 45$. It will have cut down <u>5</u> trees.
   d. $4 \times 6 = 24$. The hummingbird will feed <u>24</u> times.
   e. The two batches take 50 minutes. The muffins will be done at <u>3:20 PM</u>.

Page 32

2. a. Pinto
   b. $28 - 13 = 15$. There are <u>15</u> more bay horses.
   c. $28 + 9 + 21 + 13 + 7 - 2 - 1 = 75$. Gary now has <u>75</u> horses.

3. For the first purchase, he received $25 − $21 = $4 in change.
   For the second purchase, he received $30 − $26 = $4 in change. He received a total of <u>$8</u> in change.

Money and Change, pp. 33-35

Page 33

1. a. $2.40 + $2.40 = $4.80. Yes, you can buy two boxes. Your change is $0.20.
   b. $3.80 + $1.50 = $5.30. No, you cannot buy them; you need $0.30 more.
   c. $3.70 + $3.70 + $3.70 = $11.10. No, you cannot buy them; you need $1.10 more.

Page 34

2. a. 245 cents = $2.45. Cole had $2.45 + $5 = $7.45
   b. The total is $0.35 + $0.99 + $2.05 = $3.39
   c. Her meal cost $50 − $31.50 = $18.50.
   d. $4.35 + $4.35 + $4.35 = $13.05; $15 − $13.05 = $1.95. His change was $1.95.
   e. No. The change should be $10 − $5.23 = $4.77.
   f. $15 − $11.58 = $3.42. Yes, the change is correct.

Page 35

3. a. Total: $4.85 + $3.15 = $8.00. Change: $0.
   b. Total: $4.85 + $11.58 + $5.23 = $21.66. Change: $30 − $21.66 = $8.34.
   c. Two salads cost $8.50 + $8.50 = $17, and four cost $17 + $17 = $34.
   d. Salad and yogurt is $8.50 + $5.23 = $13.73. Soup and pizza is $11.58 + $3.15 = $14.73. Salad and yogurt is cheaper.

More Multiplication, pp. 36-37

Page 36

1. Student number sentences may include an unknown, but they don't have to.
   a. $6 \times 3 + 4 = 22$. They cost $22.
   b. $5 \times 12 + 2 \times 11 = 82$. There are 82 passengers.
   c. $7 \times d = 28$; $d = 4$. There will be 4 layers.
   d. $3 \times 12 = 36$. Mom bought 36 eggs.
   e. $2 \times p = 18$; $p = 9$. Aiden bought 9 packages.
   f. $6 \times 5 + 4 = 34$. Mom is 34 years old.

Page 37

2. a. $4 \times 6 + 5 \times 5 + 2 \times 2 = 53$. There are 53 clownfish.
   b. $7 \times 3 + 6 = 27$. The airplane cost $27.
   c. You want to spend no more than $30 − $10 = $20. $5 \times 4 = 20$, so you can buy five cans of tennis balls.

Puzzle corner: Drawing the situation can help.

First find out how many groups of four seconds Mack takes to cross the pond: $6 \text{ ft} \times s = 42 \text{ ft}$; $s = 7$. Mack takes seven groups of four seconds to cross the pond. $7 \times 4 = 28$. Mack takes 28 seconds to cross the pond.

Louie crosses the pond in groups of seven feet. Since $6 \times 7 \text{ ft} = 42 \text{ ft}$, Louie takes six jumps that are seven feet. Each of his jumps, except the last one, takes a total of four seconds (jump time and rest time). The last jump takes one second and then he is across. $5 \times 4 + 1 = 21$. Louie takes 21 seconds to cross the pond.

Louie won the race.

Mixed Word Problems 2, pp. 38-39

Page 38

1. a. Driving halfway there and then back is the same as one full trip, so Mom basically made three trips. $3 \times 30 = 90$. Mom drove <u>90 miles</u>. Or, you can add: $15 + 15 + 30 + 30 = 90$.

   b. First find how long it takes to put all the marbles on the run. The last marble doesn't count because there are no more four-second delays after putting it on. It takes $10 \times 4 = $ <u>40 seconds</u> between putting the first and last marble on the run. Then add the one minute, or 60 seconds, it takes for the last marble to finish the slides. It takes $40 + 60 = $ <u>100 seconds</u> for all the marbles to finish.

   c. $444 - 396 = 48$. There are 48 second-grade tomatoes going into bags of eight. $8 \times 6 = 48$, so he will get <u>6 bags</u>. $6 \times \$1.50 = (6 \times \$1) + (6 \times \$0.50) = \$6 + \$3 = \$9$. He will make <u>\$9</u> from the second-grade tomatoes.

Page 39

2. Students' answers will vary.

   a. EXAMPLE QUESTION: How many pages will Brooklyn read in total? Answer: $340 + 196 + 153 + 216 = $ <u>905 pages</u>. Or, how many more pages does the book with the most pages have than the book with the least pages? Answer: $340 - 153 = $ <u>187 more pages</u>.

   b. EXAMPLE QUESTION: How many balloons did Grant have in total? Answer: $7 \times 4 + 12 = $ <u>40 balloons</u>.

   c. EXAMPLE QUESTION: How many more mice did Fred catch during the first two weeks than the third week? Answer: During the first two weeks, Fred caught $14 \times 2 = 28$ mice. During the third week, Fred caught 7 mice. So, he caught $28 - 7 = $ <u>21 more mice</u> during the first two weeks than during the third week.

   d. EXAMPLE QUESTION: What do the desk and the chairs cost in total? Answer: $\$154 + \$128 = $ <u>\$282</u>. Or, Bob pays for the desk and chairs with \$300. What is his change? Answer: $\$300 - \$154 - \$128 = $ <u>\$18</u>.

3. a. 54. Guess and check: try each of the numbers from 50 to 60. The number 54 is the only one that will work, because the sum of the digits of 54 is 9, $54 - 9 = 45$, and $4 + 5 = 9$.

   b. 22. You can check out every number from 15 to 85, but as you start doing so, you will probably notice some facts.

   The product of the digit is almost always more than the sum of the digits. For example, for 64, $6 + 4 = 10$ but $6 \times 4 = 24$. That is just the nature of multiplication; it produces a bigger number. For the sum and the product to be equal, the digits need to be small. When one of the digits is zero, the product is zero, and that doesn't work. When one of the digits is 1, the product is the same as the other digit, but the sum cannot be that, so that won't work. Reasoning like this, you will soon be checking out if numbers with 2s and 3s work, and 22 is the one that fits the bill.

   c. 42. Check all the numbers in the table of 7 (7, 14, 21, 28, and so on) and you will find that 42 is the only one that works. Double 42 is 84, and $8 \times 4 = 32$.

Word Problems with Large Numbers, pp. 40-41

Page 40

1. a. $2,380 + 2,380 = 4,760$. The geese flew <u>4,760 miles</u>.

   b. $1,929 - 500 = 1,429$. The other camera cost <u>\$1,429</u>.

   c. $5,600 - 4,200 = 1,400$. The farmer got a 1,400 egg increase from seven chickens. $5,600 + 1,400 = 7,000$. The farmer can expect to get <u>7,000 eggs</u>.

   d. $6,200 + 1,400 + 140 = 7,740$. The package traveled <u>7,740 km</u>.

Page 41

2. a. The shorter track is $3,500 - 840 = 2,660$ feet. $2,660 + 2,660 = 5,320$. Tristan jogged <u>5,320 km</u>.

   b. $3,550 - 1,200 = 2,350$. The price difference is <u>\$2,350</u>.

   c. $2,890 + 2,890 + 4,150 = 9,930$. The cost was <u>\$9,930</u>.

   d. $7,590 - 6,324 + 1,500 = 2,766$. She needs <u>\$2,766</u> more.

Page 42

1. a. The perimeter is 415 ft + 415 ft + 144 ft + 144 ft = <u>1,118 feet</u>.

   b. The area is 9 ft × 12 ft = <u>108 ft^2</u>.

2. a. The area was 32 × 2 ft^2 = 64 ft^2.
   b. First find the length of the other side. The two sides equal half of the perimeter: 6 m + ? = 10 m, so the other side
      is 4 m. Now, the area is 4 m × 6 m = <u>24 m^2</u>.
   c. The two sides multiplied give the area: 7 ft × y = 42 ft^2. From that, y = 6 ft.
      Then the perimeter is 2 × 7 ft + 2 × 6 ft = <u>26 ft</u>.

Page 43

3. a. The area of the painting is 10 in × 8 in = <u>80 in^2</u>. The area of the frame and painting is 12 in × 10 in = 120 in^2.
      Then subtract the area of the painting from the frame: 120 in^2 − 80 in^2 = <u>40 in^2</u>.

   b. One rooster and 19 hens is 20 chickens. 20 × 4 m^2 = 80 m^2, so he needs at least 80 m^2 of outdoor space.
      The area of the coop is 3 m × 3 m = 9 m^2. The total area is 80 m^2 + 9 m^2 = 89 m^2.
      Milo needs a minimum of <u>89 m^2</u> for his chickens.

Puzzle corner: <u>Each side should be 5 meters.</u> To calculate area, we multiply
the two sides. Those two sides add up to half of the perimeter, which is
10 meters. You can test all the possible combinations for the two side
lengths that add up to 10, to see which one has the largest possible area.
For example, the sides 1 m and 9 m have an area of 1 × 9 = 9 m^2.

From the table we can see that a 5 m by 5 m garden has the maximum area.
If the side lengths are decimal numbers between the ones given in the
table, the area will be between the numbers in the table, and thus still
less than 25 m^2.

Side 1	Side 2	Area
1 m	9 m	9 m^2
2 m	8 m	16 m^2
3 m	7 m	21 m^2
4 m	6 m	24 m^2
5 m	5 m	25 m^2

Word Problems with Measurement Units, pp. 44-45

Page 44

1. a. Since 1 m 35 cm = 135 cm, Makayla is taller by 1 cm.
   b. The perimeter is 4 × (3 ft 8 in) = 12 ft 32 in = 14 ft 8 in
   c. Since 20 × 50 g = 1,000 g, <u>20 light bulbs</u> would weigh 1 kg.
   d. 6 × 5 oz = 30 oz = 1 lb 14 oz
   e. In eight weeks, she uses 8 × 20 ml = 160 ml. Then, 1,000 ml − 160 ml = 840 ml will be left.
   f. The total weight is 880 g + 650 g + 1,100 g = <u>2,630 g</u>
   g. 3 × 11 in + 4 × 4 in = 49 in = <u>4 ft 1 in</u>
   h. 2 m + 1 m + 28 cm = 3 m 28 cm = <u>328 cm</u>

Page 45

2. a. Half of 3 pounds is 1 lb 8 oz, or 24 oz. Then, 24 oz ÷ 3 oz = 8. There are <u>8 chocolate bars weighing 3 oz each</u>.

   b. Note that 2 liters = 2,000 ml. At the point when there was 1 L 100 ml, or 1,100 ml left, the family had drunk
      2,000 ml − 1,100 ml = 900 ml. Then we add to that the 250 ml that Roger drank: 900 ml + 250 ml = <u>1,150 ml</u>.
      So, the family has drunk <u>1,150 ml of juice</u>.

   c. The tortoise is traveling 23 + 23 + 23 = 69 feet. 69 ft − 35 ft = 34 ft. The tortoise has <u>34 feet</u> to go.

Puzzle corner: Half a gallon is two quarts, which is four pints, which is eight cups. Each time they eat ice cream, they
consume two cups (one cup for Dad, half a cup for Alana, and half a cup for Ella). 8 cups ÷ 2 cups = 4, so they can eat
ice cream four times. They have it once every two days. Let's say they have it on days 1, 3, 5, and 7. So, the ice cream will
last <u>seven days</u>.

Division in Word Problems, pp. 46-47

Page 46

As usual, student number sentences may include an unknown, but they don't have to.

1. a. $4 \times 4 = 16$ and $16 \div 4 = 4$

   b. $8 \times 6 = 48$ (or $6 \times 8 = 48$) and $48 \div 8 = 6$ (or $48 \div 6 = 8$)

   c. $9 \times 12 = 108$ (or $12 \times 9 = 108$) and $108 \div 9 = 12$ (or $108 \div 12 = 9$)

2. a. $5 \times 12 = 60$, or $60 \div 12 = 5$. There were _60_ apple slices.

   b. $11 \times 6 = 66$, or $66 \div 11 = 6$. There are _6_ in each group.

   c. $3 \times 6 = 18$, or $18 \div 3 = 6$. Each got _6_ marbles.

Page 47

As before, student number sentences may or may not have an unknown.

3. a. $18 \div 3 = 6$ or $3 \times 6 = 18$. Each mug cost $\underline{\$6}$.

   b. There are $64 \div 8 = 8$ cookies in each box. Kelly has $8 - 6 = 2$ boxes left. She has $2 \times 8 = \underline{16\ cookies}$ left.

   c. Hunter bought $27 \div 3 = 9$ energy bars. $9 - x = 6$; $x = 3$. He ate _3 bars_.

   d. $3 \times 6 = 18$. There are 18 pizza slices. $18 \div 5 = 3$ R3. So, each kid got three slices and the dog got _three slices_.

   e. $9 \times t = 180$; $t = 20$. It will take _20 minutes_ to fill.

   f. There are several ways to solve this. One is just plain guess and check. Or, you can think this way. First remove the extra marbles Dill and Will have from the situation: $30 - 3 - 3 = 24$. Then divide the rest evenly: $24 \div 4 = 6$. Will and Phil each have _6 marbles_ and Bill and Dill each have _9 marbles_.

Mixed Word Problems 3, pp. 48-49

Page 48

1. a. The tortoise has exactly one hour to eat cucumbers. One hour is 60 minutes, which is 20×3 minutes. The tortoise eats one ounce in each of those 3 minutes, so, he eats 20 ounces of cucumber by 5:30 PM.

   Since 20 oz = 13 oz + 7 oz, the tortoise will have only eaten one whole cucumber.

   b. From 12:02 to 12:15 is 13 minutes. From 4:43 to 4:59 is 16 minutes. The second crowing session is longer.

   To find the elapsed time from 12:15 AM to 4:43 AM, subtract the minutes: $43 - 15 = 28$ minutes. The hours go from 12 AM to 4 AM, which is four hours. So, he spent _4 hours and 28 minutes_ between the two sessions.

   c. Two months ago, Pudding knew $2 \times 12 = \underline{24\ words}$. Now, Pudding knows $2 \times 24 = \underline{48\ words}$.

   Since he doubled how many words he knows, he learned 24 words. Sushi learned $24 + 1 = 25$ words.

   So, now Sushi knows $12 + 25 = \underline{37\ words}$.

Page 49

2. The yellow area can be split into sections, the top portion and the lower portion. The top portion is $9\ m \times 3\ m$ $= 27\ m^2$, and the bottom portion is $3\ m \times 12\ m = 36\ m^2$. $36\ m^2 + 27\ m^2 = 63\ m^2$. The yellow area is $\underline{63\ m^2}$.

   OR: The area of the whole barn is $9 \times 15 = 135\ m^2$, and the area of each of the six stalls is $3 \times 4 = 12\ m^2$. so the area of six stalls is $6 \times 12 = 72\ m^2$. Now we subtract the area of the stalls from the total area of the barn to get the yellow area: $135\ m^2 - 72\ m^2 = 63\ m^2$. The yellow area is $\underline{63\ m^2}$.

3. The student's answers will vary.

   a. EXAMPLE ANSWER: Mom's family is crazy about eating omelets, so she buys three dozen eggs, although she already has 20 at home. How many eggs does she have now? Answer: 56

   b. EXAMPLE ANSWER: Michael had $50, and then he bought three toy dump trucks for $8 each. How much does he have now? Answer: $26

Page 50

1. a. $11 \times 5 = 55$ or $55 \div 5 = 11$. She would get <u>11 bouquets</u>.
   b. $39 \div 12 = 3$ R3. Milo got <u>three cartons</u>, and <u>three eggs didn't fit</u>.
   c. $9 \times 4 = \underline{36 \text{ people}}$ would fit.
   d. $8 \times 8 = 64$ or $64 \div 8 = 8$. There are <u>8 stickers</u> on each page.
   e. Since you use 1 cup of oats, you are making 1/3 of the recipe. For that, you will use $15 \div 3 = \underline{5 \text{ apples}}$.

2. The student's answers will vary.

   a. EXAMPLE ANSWER: The smallest dog gets two bones, and the rest each get four. How many dogs are there? Answer: The other dogs get $18 - 2 = 16$ bones. Since each of them gets four, there are $16 \div 4 = 4$ dogs, besides the smallest dog. So, all in all there are <u>five dogs</u>.

   b. EXAMPLE ANSWER: If Milo gave each his chickens one cucumber, ate 11 himself, and gave each of his neighbors 5 cucumbers each, how many neighbors got cucumbers?

   Answer: He gives $67 - 11 - 6 = 50$ cucumbers to neighbors. Since each neighbor gets 5 cucumbers, there are $50 \div 5 = \underline{10 \text{ neighbors}}$.

Page 51

3. a. The panda eats $21 \div 7 = 3$ pounds of bamboo in one hour. The panda eats $3 \times 5 = \underline{15 \text{ pounds}}$ of bamboo in five hours.

   b. There are $3 \times 6 = 18$ people in three of the speedboats. There are $63 - 18 = 45$ people in the other nine. There are $45 \div 9 = \underline{5 \text{ people}}$ in each of the other speedboats.

   c. $\$2 \times t = \$80$; $t = 40$, so Morris bought 40 tuna cans. $\$3 \times s = \$60$; $s = 20$, so Morris bought 20 sardine cans. Morris bought $40 + 20 = \underline{60 \text{ cans}}$ of fish.

Puzzle corner: Matthew had $\$125 + \$95 = \$220$ before he bought the watch. He earned $4 \times \$40 = \$160$ from other yard work, which means he earned $\$220 - \$160 = \$60$ from mowing the lawn. Matthew mowed the lawn $\$60 \div \$12 = \underline{5 \text{ times}}$.

Fraction Problems, pp. 52-53

Page 52

1. a. Johnny (because 1/3 is greater than 1/4)
   b. Mom (because 7/8 is greater than 2/10)
   c. Of the flowers, $15 - 7 - 5 = 3$ are dahlias. So, <u>3/15</u> (or 1/5) of the flowers are dahlias. Miranda has $15 \div 3 = 5$ daffodils. They have the <u>same amount</u> of daffodils.
   d. Since $4/10 = 2/5$, Max should eat <u>4 slices</u> to eat the same amount as Frank.
   e. Xavier ate the same amount, because $2/12 = 1/6$.

 Gerald

 Hank

 Xavier

 Ricardo

Page 53

2. a. You ate more. 3/4 of a BIG plateful is a larger amount than 3/4 of a small plateful.
   b. 1/4
   c. Abby ate more. She ate $3 \times 6 + 1 = 19$ sixths, which is two more sixths than Susanna.
   d. Barry ate more. He ate $2 \times 4 + 3 = 11$ fourths, which is two more fourths than Andy.
   e. $12 \div 3 = 4$. Four were green. $12 \div 2 = 6$. Six were blue. $12 \div 6 = 2$. Two were yellow.
   f. (i.) Eddy ate more. The pizzas are the same size, and 1/4 is a bigger fraction than 1/6.
      (ii.) You cannot easily tell who got more because the pizzas are different sizes.

Word Problems with Large Numbers 2, pp. 54-55

Page 54

1. a. 2,852 + 412 − 238 = 3,026. Earlier it was flying at <u>3,026 feet</u>.

   b. Estimations may vary.
      For example: $9,600 − $2,000 = $7,600. The actual price difference: $9,590 − $1,955 = <u>$7,635</u>.

   c. Estimations may vary. For example: $640 + $640 + $400 + $400 + $400 = $1,280 + $1,200 = $2,480.
      The actual cost: $637 + $637 + $395 + $395 + $395 = <u>$2,459</u>.

   d. The price for the fence posts and rails was $1,784 + $1,697 = $3,481. After the discount, the price was
      $3,481 − $499 = $2,982. Then, if c was her change, 2,982 + c = 3,000. Willow's change was <u>$18</u>.

Page 55

2. a. The giraffe is about 2,500 lb − 1,100 lb = <u>1,400 pounds</u> heavier.

   b. The elephant is about 8,000 lb − 2,500 lb = <u>5,500 pounds</u> heavier.

   c. Two hippos weigh 3,200 lb + 3,200 lb = 6,400 lb, which is not enough. Three hippos weigh
      3,200 lb + 3,200 lb + 3,200 lb = 9,600 lb, which is more than the elephant's weight. You would need <u>three hippos</u>.

   d. The <u>bear and rhino</u> (700 lb + 2,900 lb = 3,600 lb) and the <u>horse and giraffe</u> (1,100 lb + 2,500 lb = 3,600 lb) have
      the same combined weight.

Puzzle corner: There were two adult penguins for each two eggs, which means there was the same number of eggs as adults. Combine the eggs that did hatch (the chicks) and the ones that didn't: 4,241 + 231 = 4,472. This means there are 4,472 adults. Then, there are 4,472 + 4,241 = <u>8,713 penguins</u> in total on the island.

Mixed Word Problems 4, pp. 56-57

Page 56

1. a. Oliver found 2 × 9 = 18 shells. Marlene found 44 − 18 − 9 = <u>17 shells</u>.

   b. The total cost of the glitter pens is 3 × $11 = <u>$33</u>.

   c. Clara can still eat 20 g − 8 g = 12 g of sugar today. Each cookie has 15 g ÷ 3 = 5 grams of sugar.
      She can eat <u>two whole cookies</u>.

   d. Marvin read 2 + 4 + 2 + 4 + 2 + 4 + 2 = <u>20 short stories</u>.

Page 57

2. The perimeter of the yard is 24 + 24 + 30 + 30 = 108 feet. The bones will be 108 ÷ 12 = <u>9 feet apart</u>.

3. Student answers will vary. EXAMPLE ANSWER: Martha uses 9 strawberries a day in her smoothie. Strawberries are
   sold in packages of 24. How many packages will she need to buy to last her a week of smoothies?

   Answer: She will make seven smoothies, so she needs 7 × 9 = 63 strawberries. Two packages have 24 + 24 = 48
   strawberries, so that is not enough. Three packages have 24 + 24 + 24 = 72 strawberries, which is more than 63,
   so <u>three packages is enough</u>.

Puzzle corner: He gave her <u>3 cents (or $0.03)</u>. After receiving the initial change, Luca had paid $10 − $2.20 = $7.80, which was 3 cents less than the price of the car.

More from math MAMMOTH

Math Mammoth has a variety of resources to fit your needs. All are available as economical downloads, and most also as printed copies.

- **Math Mammoth Light Blue Series**
  A complete curriculum for grades 1-7. Each grade level includes two student worktexts (A and B), which contain all the instruction and exercises all in the same book, answer keys, tests, cumulative reviews, and a worksheet maker. International (all metric), Canadian, and South African versions are also available.

  https://www.MathMammoth.com/complete-curriculum

  https://www.MathMammoth.com/international/international

  https://www.MathMammoth.com/canada/

  https://www.MathMammoth.com/south_africa/

- **Math Mammoth Skills Review Workbooks**
  These workbooks are intended to be used alongside the Light Blue series full curriculum, and they provide additional review to the topics studied in the main curriculum, in a spiral manner.

  https://www.MathMammoth.com/skills_review_workbooks/

- **Math Mammoth Blue Series**
  Blue Series books are topical worktexts for grades 1-8, containing both instruction and exercises. They cover all elementary math topics from 1st through 7th grade and some for 8th grade. These books are not tied to grade levels, and are thus great for filling in gaps.

  https://www.MathMammoth.com/blue-series

- **Make It Real Learning**
  These activity workbooks concentrate on answering the question, "Where is math used in real life?" The series includes various workbooks for grades 3-12.

  https://www.MathMammoth.com/worksheets/mirl/

- **Review Workbooks**
  Workbooks for grades 1-7 that provide a comprehensive review of one grade level of math—for example, for review during school break or summer vacation.

  https://www.MathMammoth.com/review_workbooks/

Free gift!

- Receive over 350 free sample pages and worksheets from my books, plus other freebies:
  https://www.MathMammoth.com/worksheets/free

Lastly...

- Inspire4 is an inspirational website for the whole family I've been privileged to help with:
  https://www.inspire4.com